OFF TO SPACE!

STARS

by Rachel Rose

Consultant: Beth Gambro
Reading Specialist, Yorkville, Illinois

BEARPORT PUBLISHING

Minneapolis, Minnesota

Teaching Tips

Before Reading

- Look at the cover of the book. Discuss the picture and the title.
- Ask readers to brainstorm a list of what they already know about stars. What can they expect to see in this book?
- Go on a picture walk, looking through the pictures to discuss vocabulary and make predictions about the text.

During Reading

- Read for purpose. Encourage readers to think about stars and the role they play with Earth and space as they are reading.
- If readers encounter an unknown word, ask them to look at the sounds in the word. Then, ask them to look at the rest of the page. Are there any clues to help them understand?

After Reading

- Encourage readers to pick a buddy and reread the book together.
- How is a star born and how does it die? Find the pages that tell about these things.
- Ask readers to write a sentence and draw a picture of something that they learned about stars.

Credits:
Cover and title page, © NASA/JPL; 5, © mik38/iStock; 7, © ferrantraite/iStock; 9, © Triff/Shutterstock; 10, © NASA Goddard; 12-13, © NASA/JPL; 15, © aryos/Shutterstock; 16–17, © shooarts/Shutterstock; 19, © tose/iStock; 21, © Veronika_Sh_art/Shutterstock; 22, © Artsiom Petrushenka/Shutterstock; 23BL, © Vladimir Sazonov/Shutterstock; 23BM, © NASA Goddard; 23BR, © melnyk mariya/Shutterstock; 23TL, © NASA/JPL; 23TR, © Martin Capek/Shutterstock

Library of Congress Cataloging-in-Publication Data is available at www.loc.gov or upon request from the publisher.

ISBN: 978-1-64747-508-6 (hardcover)
ISBN: 978-1-64747-513-0 (paperback)
ISBN: 978-1-64747-518-5 (ebook)

For more information, write to Bearport Publishing, 5357 Penn Avenue South, Minneapolis, MN 55419.

Printed in the United States of America.

Contents

Twinkle, Twinkle, Little Star!

The night sky is dark.

Look up and see the stars shine.

The darker it is, the more stars you can see!

There are thousands of stars in our sky.

But there are many more stars we do not see.

Stars look small because they are far away.

But stars are very big.

The closest star to us is the sun.

10

Stars are huge balls of burning **gas**.

The gas gives off heat and light.

A star starts in a cloud of **dust** and gas.

The gas and dust **form** a ball.

The ball spins and gets very hot.

Then, a star is born.

Stars can be different colors.

Their colors show how hot they are.

The hottest stars are blue.

The coldest ones are red.

Stars can be different sizes, too.

They can get bigger or smaller over time.

Some stars start small.

Then, they get really big.

Stars can live for billions of years.

They die when they run out of gas.

Some stars **explode** before they die.

But much of this happens far from Earth.

Closer to home, what can you see?

Sometimes stars make shapes in the sky.

Can you spot a bear?

Look up at the stars!

Fun Star Facts

There are more stars in space than **grains** of sand on Earth.

Light from faraway stars takes years to make it to Earth. You are looking into the past when you see it!

Some lights in the sky shine like stars. But they are really things humans put in space.

Glossary

dust a fine powder

explode to suddenly break apart

form to take shape

gas a form with no shape, such as air

grains small, hard pieces of something

Index

Read More

Moon, Walt K. *Let's Explore the Stars (A First Look at Space).* Minneapolis: Lerner Publications, 2018.

Rector, Rebecca Kraft. *Stars (Let's Learn About Space).* New York: Enslow Publishing, 2019.

Learn More Online

1. Go to **www.factsurfer.com**
2. Enter "**Space Stars**" into the search box.
3. Click on the cover of this book to see a list of websites.

About the Author

Rachel Rose spends many nights looking at the stars. And making wishes.